MW00635016

THE

TYRRELLS OR TERRELLS

OF

AMERICA.

Tyrrell of Thornton.

The Genealogy of

Richmond and William Tyrrell
or Terrell

(Descended from the family of Tyrrell of
Thornton Hall, Buckinghamshire, England),

Who settled in Virginia in the

Seventeenth Century.

COMPILED

(And issued as a Supplement to his "Genealogical
History of the Tyrrells") by

JOSEPH HENRY TYRRELL.

150 COPIES PRIVATELY PRINTED. *No.*........ ....

FOREWORD.

EMBOLDENED by the approval of my Kinsmen of my attempt at the " Genealogical History of the Tyrrells," I have thought that a Supplement to that work, dealing with the founding of the now widely spread American family, may meet with favour.

The difficulties that stood in the way of discovering the lineage of Richmond and William, the original emigrants, were very great, and at times I should have been tempted to relinquish the search but for the encouragement received at the hands of a distinguished member of the American branch—The late Honourable Edwin Holland Terrell.

This work is not in any sense intended to be a complete record of the family, but should be read in conjunction with the above-named history.

J. H. T.

CASTLEKNOCK,

TWICKENHAM, *October*, 1910.

2

CONTENTS.

ARMS OF TYRRELL OF THORNTON.

Argent, two chevrons, azure, within a bordure, engrailed, gules: a crescent, sable, for difference.

CREST: A boar's head erect and couped argent, tusked or: from the mouth a peacock's tail issuant, proper.

MOTTO: " Sans dieu rien."

BADGE: A triangular curved fret, or.

ATCHIEVEMENT OF TYRRELL OF THORNTON.

1. Tyrrell of Thornton.
2. Vexin.
3. Tirel (*ancient*).
4. „ of Poix.
5. Tyrrell of Avon Tyrrell.
6. „ of Castleknock.
7. „ of Heron.
8. Dreux.
9. Guernanville.
10. Giffard.
11. Burgate.
12. Flambert.
13. Heron.
14. Swynford.
15. Coggeshall.
16. Staunton.
17. Watteville.
18. Hawkwood.
19. Le Brun.
20. Rokeley.
21. Radford.
22. Bodley.
23. Ingleton.
24. Fitzellis.
25. Barton.
26. Tyrrell of Thornton.

KEY CHART TO ATCHIEVEMENT.

(1—26) Tyrrell of Thornton, descended from

(2) Walter I., Count of Vexin. = Eve, dau. and heiress of Landry, Count of Dreux (8)

(8) Ralf, Sire de Tirel. = Dau. and heiress of de Guernanville (9).
a quo.

(4) Sir Walter, of Poix. = Olga.
a quo

Sir Walter, of Poix. = Adelaide, dau. and heiress of Walter Giffard (10)

(5) Sir Hugh, of Avon. = Ada, dau. of Etienne, Count d'Aumale.

(6) Sir Hugh, of Castleknock, = Marie de Senarpont.
a quo.

Sir Edward. = Jane, dau. and heiress of Sir Wm. Burgate (11).

Sir Hugh. = Jane, dau. and heiress of Sir James Flambert (12).

Sir John de Coggeshall, = Mabel, sis and heiress of Anfrus de Staunton (16).
a quo.

(7) Sir James. = Margaret, dau. and heiress of Sir Wm. Heron (13).

Richard. = Dau. and heiress of Sir John Watteville (17).

Sir Morice Le Brun, = Isolde, dau. and heiress of Philip de Rokeley (20).
a quo.

Sir Walter, = Anna, dau. and heiress of Sir John Swynford (14).
a quo.

Sir William. = Mary, dau. and heiress of Sir John Hawkwood (18)

Sir Morice. = Elizabeth, dau. and heiress of Sir Henry Radford (21).

Sir John, = Eleanor, or another, dau and heiress (15).
a quo.

Sir Humphrey. = Elizabeth, dau. of Robert D'Arcy.

Sir Thomas. = Elizabeth, dau. and heiress (19).

Ingleton. = Dau. and heiress of —— Fiteells (24).

William. = Elizabeth, dau. and heiress of Thomas Bodley (22).

John Ingleton. = Dau. and heiress of Barton (25).

Humphrey. = Jane, dau. and heiress (28).

BIOGRAPHICAL · NOTE.

The early history of the Tyrrells is extensively set out in the "Genealogical History of the Tyrrells," from which work the first portion of the following sketch is extracted.

The Tyrrells of Thornton Hall, 4 miles N.E. of Buckingham, were descended from Sir James Tyrrell, who lived about the middle of the fourteenth century, and who married Margaret, daughter and heiress of Sir William Heron, of Heron Hall, Essex.

This Sir James was a direct descendant of Sir Walter, the alleged Regicide, whose grandfather, Sir Walter, was one of the companions of the Conqueror in 1066; and who was descended from Pepin le Gros, Duke of Brabant, great grandfather of Charlemagne.

Sir James was great grandfather to Sir John of Heron, who was present at Agincourt, 1415; Sheriff of Essex, 1423; Treasurer of the Household to Henry VI; several times Speaker of the House of Commons; and who died in 1437. His son, Sir Thomas, was Chamberlain of the Exchequer, and died in 1476, leaving four sons, the second of whom, Sir Thomas of Ockendon, was father of William of Ockendon, who married Elizabeth, daughter, and eventual heiress of Thomas Bodley of Devon. They had two sons, the second of whom, Humphrey, married Jane, daughter and heiress of John Ingleton of Thornton Hall, and obtained with her a very large estate in Buckinghamshire and Oxfordshire, comprising over thirty manors. She was a Ward of the Thomas Bodley above named. They had one son, George, who married Elizabeth, or Eleanor, daughter of Sir Edward Montagu, Lord Chief Justice of England. George Tyrrell died on 16 May, 1571, leaving Sir Edward, his heir; William, of whom later; Thomas of London; Francis of London; and three daughters. Sir Edward married and had issue: Henry, who died young; Edward, created a Baronet in 1627; Sir Timothy; Francis; Charles (see Will, 1619); John; Sir Thomas; and eight daughters. The descendants of Sir Edward carried on the title until 1749, when the last Baronet, Sir Charles, died.

Reverting to William, second son of George, he is named in " Morant's History of Essex " as being siezed of part of Ockendon, and of the Manor of Bruyn or Brun, in Essex, which, in course of time, he alienated. He settled at Reading, not many miles from Thornton, and is variously described as Tyrrell, Terrell, Tirrell and Terrill, and his Virginian descendants have, in the main, adopted the second mode of spelling the name: possibly owing to that form having been most frequently used in the early Land Grants. He married a daughter of ....
Richmond *alias* Webb, of Stewley, Bucks, and had issue amongst others: Robert of Reading, who married Jane, daughter of Robert

8

Baldwin, at St. Giles' Church, Reading, on 29 June, 1617.   They had
issue: John of Reading (Will, 1661), Robert of London (Will, 1677),
Richmond, William, and Mary.

John and Robert died without issue.   Richmond and William settled
in Virginia and became the Ancestors of all the Tyrrells, Terrells, Tirrells
and Terrills of America, with the exception of some few descended
from the Irish branch of the family.

John, the eldest son, seems to have lived and died at Reading.
Robert was a merchant in London, with interests in Virginia, as appears
by his Will, 1677.   He was a Witness to a deed in York County,
24 January, 1647, and in the records of that county there is enrolled
a power of Attorney from John Wiskens and John Robinson, merchants,
of London, to Robert Terrell, Citizen and Fishmonger, of London, to
collect all debts due to them in Virginia.

In the same County is also enrolled a power of Attorney from
Robert Terrell, Citizen and Fishmonger, of London, intending to take a
voyage to England, appointing Thomas Williamson his Attorney in
Virginia.

Richmond and William settled in Virginia, the former prior to
1650 and the latter before 1657.   They and their descendants received

sundry grants of lands, as appears by the following extracts from the Colonial records. From them descend the various branches now resident in the United States.

LAND GRANTS.

Richmond Terrell, 640 acres in New Kent on the S.W. side of York River.—23 November, 1658.

Richmond Tirrell, 600 acres on Chickahominy River in New Kent. —8 February, 1670.

Robert Tirrell, 170 acres in St. Stephen's Parish, New Kent, on the north side of Mattapony River.—20 April, 1682.

Mrs. Elizabeth Terrell and Thomas Correll, 720 acres in New Kent, on the south side of York River.—20 November, 1683.

Robert Terrell, 63 acres in Middlesex, on a branch of Parrott's Creek.—24 October, 1701.

William Terrell, 300 acres on the S.E. side of Pole Cat Swamp, King William County.—16 June, 1714.

William Terrell, 100 acres on the N. side of Pamunkey, in King William County.—22 March, 1715.

William Terrell, 400 acres on a fork of Pole Cat Swamp.—31 October, 1716.

William Terrell and Robert Chandler, 300 acres on a fork of Pole Cat Swamp.—1 April, 1717.

William Terrell and his son William, 400 acres on Pole Cat Swamp. —18 March, 1718.

William Tyrrell, Jun., 174 acres on Pole Cat Swamp.—12 July, 1718.

William Terrell, of New Kent, 400 acres on Pole Cat Swamp.— 22 January, 1718.

Joel Terrell, of King William, 400 acres in St. Margaret's Parish, King William.—9 July, 1724.

William Terrell, Sen., of New Kent, 220 acres on Pole Cat Swamp —22 January, 1718.

William Terrell, of King William, 237 acres on Pole Cat Swamp, St. Margaret's Parish.—22 February, 1724.

William and Susanna Terrell, by deed dated 16 March, 1725, conveyed 400 acres of land in King William County, to their son Henry.

Joseph Terrell, of New Kent, 400 acres in Hanover County.— 7 September, 1729.

John Terrell, of Caroline County, 400 acres in Spotsylvania County, on the north side of Rapidan River.—28 September, 1730.

Richmond Terrell, 450 acres in Hanover.—12 April, 1732.

Richmond Terrell, of New Kent, by deed dated 29 April, 1670, conveyed 600 acres to Henry Wyatt. The deed reserves " 100 acres formerly given unto my brother William Terrell, and since sold by him to Francis Waring."

———

Owing to the destruction of many of the records of New Kent, King and Queen, King William, and Hanover Counties, much information regarding the family has been lost. The descendants of Richmond Terrell lived in Louisa, Hanover, &c, Counties, and some notices of them are given in the *William and Mary Quarterly*, xiii. 263-5. The Registers of St. Peter's, New Kent, contain entries of the baptism of two sons of Timothy Terrell, viz. :—Robert, on 25 December, 1697 ; and Joseph, 31 December, 1699.

The descendants of William Terrell are much more numerous than those of Richmond. Many of the members of this branch became Quakers, and from the records kept by the Society of Friends much information may be obtained from about the year 1735. (See Bell's " *Our Quaker Friends.*")

There are some descendants of Richmond Terrell in Virginia and other southern States to this day. William, his brother, settled in St. Paul's Parish, Hanover County, Virginia ; both he and his wife were members of the Established (Episcopal) Church.

Some considerable information as to the American family may be gleaned from a pamphlet compiled, and issued, in 1909 by the late Hon. Edwin H. Terrell, of San Antonio, Texas.

WILLS.

TYRRELL.

1600.—THOMAS TYRRELL of London, Citizen and Grocer.

Property in East and West Ham, Essex, and at Mitcham and Croydon, in Surrey, to son Thomas.

Property in St. Nicholas Cole Abbey, St. Christopher-le-Stocks, and St. Ethelburga, London, to wife Margaret for life, then to son Thomas.

Property in Crawley, Surrey, and personal estate into three parts—one share to wife, and one share each to daughters Lucy and Frances.

£100 to daughter Elizabeth, wife of Gamaliel Crump ; 20 nobles to mother (Elizabeth), widow.

A country house at Woodford mentioned.

Legacy to poor of Woodford.

„ „ Grocers' Company of London.

„ „ friend, Master Robert Wright (Minister).

Wife Margaret and son Thomas executors. Probate by both.

1620.—CHARLES TYRRELL, of St. Andrews-by-the-Wardrobe, London
(dated 8 March, 1619).

To be buried in St. Andrews with wife.

Cousin Thomas Lea
Brother Thomas
    „   Timothy
Brother-in-law Gardner
Sister Penelope Gardner
Cousin William Trye
    „   Edmond West
Sister Bridget
    „   Jane
} All named in Will.

Sir Edward Tyrrell, Kt.
Cousin Robert Tyrrell
} Overseers of Will.

1609.—FRANCIS TYRRELL, of London, Citizen and Grocer.

Sister Emma (Mrs. Thaire)
    „ (Mrs. Weekes) and her son and his wife and children
Niece Alice Weeks, a minor
Cousins Thomas Tyrrell (son of my brother), Elizabeth and Frances
Various Godchildren
} All named in Will.

1638.—FRANCIS TYRROLL, of Reading, Broadweaver.

> To son Francis, fee simple of house in occupation of Isaac Yorke.
>
> To wife Elizabeth, rights of dower.
>
> To son Nicholas, a house.
>
> „ John, £60.
>
> „ Robert, £60.
>
> To daughter Elizabeth, £60.
>
> Overseers of Will : Brother Robert and brother-in-law Adam Kerbie.

1639.—THOMAS TYRRELL, of Fulborne, Cambs., Esquire.

> Lands in Fulborne, Croydon, Mitcham, St. Nicholas Cole Abbey, London.
>
> Wife Joan.
>
> Grandchild Thomas Tyrrell, a minor.
>
> Grand-daughter Elizabeth Tyrrell.
>
> Sister Mrs. Cruys.

1661.—JOHN TYRRELL, of Reading, Clothier.

> Property at Inholmes, Berks, and Eversley, Hants.
>
> Brother Robert and his heirs to have Estate
>
> Uncle Baldwin
>
> Brother Richmond
>
> „ William } Named in Will.
>
> Sister Mary Mewe
>
> Cousin Robert Tyrrell, of Reading, Maltster
>
> Proved 1672.

1677.—ROBERT TYRRELL, of London, Merchant.

> To nephew William, son of brother William, £10.
>
> To niece ...., daughter of brother William, £5.
>
> To brother Richmond, 10s. for a ring; silver cup and three spoons.
>
> To sister Mary Mewe, silver cup.
>
> Lands and Accounts in Virginia, Lands in Hants, Property in St. Nicholas Cole Abbey, London.

ARRIVALS IN VIRGINIA.

(Ex Records of Lands Bought from the Crown.)

Thomas Terrell, immigrant,		18 May, 1637.
James Tyrrill	„	12 June, 1648.
Katherine Terrell	„	10 January, 1649.
Edward. „	„	and purchaser, 28 November, 1656.
Robert „	„	28 November, 1656.
Edward „	„	10 July, 1658.
Richard Terrill	„	12 November, 1666.
Hen. Tirrell	„	20 September, 1667.
Robert „	„	20 September, 1667.
Hen. „	„	20 September, 1667.
William „	„	20 September, 1668.
Ann Terrill	„	31 May, 1678.
Tho. „	„	20 November, 1678.
William Tyrrell	„	27 February, 1670.

ROYAL DESCENT.

Note.—As descendants of Edward I. and Eleanor of Castile, it follows that the Tyrrells or Terrells of Essex, Bucks, Berks and of America, claim descent from the Saxon Kings of England, the Kings of France and Spain, the Dukes of Normandy, the Counts of Flanders, &c.

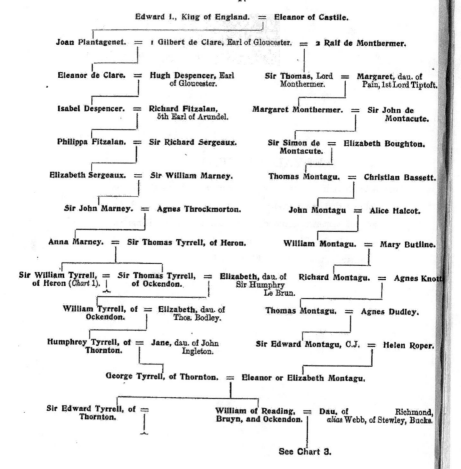

Edward I., King of England. = Eleanor of Castile.

Joan Plantagenet. = 1 Gilbert de Clare, Earl of Gloucester. = 2 Ralf de Monthermer.

Eleanor de Clare. = Hugh Despencer, Earl of Gloucester.

Sir Thomas, Lord Monthermer. = Margaret, dau. of Pain, 1st Lord Tiptoft.

Isabel Despencer. = Richard Fitzalan, 5th Earl of Arundel.

Margaret Monthermer. = Sir John de Montacute.

Philippa Fitzalan. = Sir Richard Sergeaux.

Sir Simon de Montacute. = Elizabeth Boughton.

Elizabeth Sergeaux. = Sir William Marney.

Thomas Montagu. = Christian Bassett.

Sir John Marney. = Agnes Throckmorton.

John Montagu = Alice Halcot.

Anna Marney. = Sir Thomas Tyrrell, of Heron.

William Montagu. = Mary Butline.

Sir William Tyrrell, of Heron (Chart 1). = Sir Thomas Tyrrell, of Ockendon. = Elizabeth, dau. of Sir Humphry Le Brun.

Richard Montagu. = Agnes Knott

William Tyrrell, of Ockendon. = Elizabeth, dau. of Thos. Bodley.

Thomas Montagu. = Agnes Dudley.

Humphrey Tyrrell, of Thornton. = Jane, dau. of John Ingleton.

Sir Edward Montagu, C.J. = Helen Roper.

George Tyrrell, of Thornton. = Eleanor or Elizabeth Montagu.

Sir Edward Tyrrell, of Thornton. =

William of Reading, Bruyn, and Ockendon. = Dau. of alias Webb, of Stewley, Bucks. Richmond,

See Chart 3.

CHART 1. # PEDIGREE OF TYRRELL OF HERON, ESSEX.

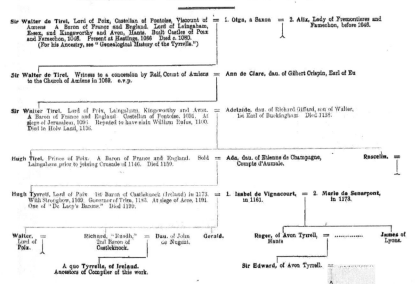

Sir Walter de Tirel, Lord of Poix, Castellan of Pontoise, Viscount of Amiens A Baron of France and England. Lord of Laingaham, Essex, and Kingsworthy and Avon, Hants. Built Castles of Poix and Famechon, 1046. Present at Hastings, 1066 Died c. 1080. (For his Ancestry, see " Genealogical History of the Tyrrells.") **= 1. Olga, a Saxon = 2. Alix, Lady of Fremontieres and Famechon, before 1046.**

Sir Walter de Tirel. Witness to a concession by Ralf, Count of Amiens to the Church of Amiens in 1069. o.v.p. **= Ann de Clare,** dau. of Gilbert Crispin, Earl of Eu

Sir Walter Tirel. Lord of Poix, Laingaham, Kingsworthy and Avon. A Baron of France and England Castellan of Pontoise, 1091. At siege of Jerusalem, 1099. Reputed to have slain William Rufus, 1100. Died in Holy Land, 1136. **= Adelaide,** dau. of Richard Giffard, son of Walter, 1st Earl of Buckingham Died 1138.

Hugh Tirel, Prince of Poix. A Baron of France and England. Laingahara prior to joining Crusade of 1146. Died 1150. **Sold = Ada,** dau. of Etienne de Champagne, Compte d'Aumale. **Roscelin. =**

Hugh Tyrrell, Lord of Poix 1st Baron of Castleknock (Ireland) in 1173. With Strongbow, 1169. Governor of Trim, 1183. At siege of Acre, 1191. One of " De Lacy's Barons." Died 1199. **= 1. Isabel de Vignacourt, = 2. Marie de Senarpont,** in 1161. in 1173.

Walter, Lord of Poix. **= Richard, "Rusdh," =** 2nd Baron of Castleknock. **Dau. of John de Nugent. Gerald. Roger,** of Avon Tyrrell, **= ....** **James of** Hants Lyons.

A quo Tyrrells, of Ireland. Ancestors of Compiler of this work.

Sir Edward, of Avon Tyrrell. **= ....**

A

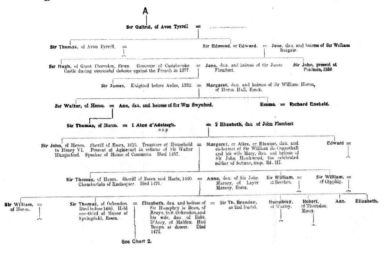

See Chart 2.

PEDIGREE OF TYRRELL OF THORNTON HALL, BUCKINGHAMSHIRE

CHART 2.

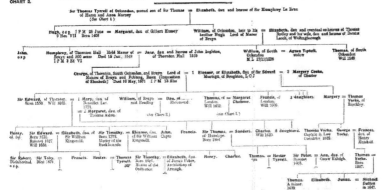

PEDIGREE OF TYRRELLS OF READING, BERKS

CHART 3.

George Tyrrell of Thornton. = Eleanor or Elizabeth, dau of Sir Edward Montage, L.C.J
(Chart 2)

| Sir Edward, of Thornton. = | William, Lord of Manor = of Denye, which he alienated, and settled at Reading. (Chart 2.) | Daughter of = of Merkley, Bucks. | Redmond. | Thomas, of = London. Will 1603. | Margaret Chabour. | Francis, of London. Will 1606. | Emma = Thure, | Dau. = Weekes. | Hester. = Sir Tim Salusbury. |

| Robert, Councillor of = Reading Borough Guardian, 1616 to, Giles Ward, 1628. | Jane, dau. of Robert Baldwin at St. Giles, Reading, 29 June, 1617. Died 1680. Her mother, Joane Pipston, was married to Robert Baldwin at St. Mary's, Reading, 5 October, 1595. | David, Councillor of Bristol. | Thomas. = | | Francis, Councillor = of Reading, 1632. Will 1638. | Elizabeth Kenby. | Thomas, of = Jane, Pinbourne, Corrie. Will 1685. | Elizabeth. | Lucy. | Frances. |
| | | | Nicholas. = Margaret. 1616-80. | | | | | | | |

| John, of Reading Will 1694. Born 25 June, 1618. | Robert, of London. Will 1677. Born 14 Nov., 1619. | Mary = John Mewe. Born 1620. | Richmond, settled in Virginia. Grant of lands in New County Kent, 1698. (Chart 4.) | William, = Martha settled in New Kent County, Virginia, 1667. | Robert, of Reading = Bennett. Born 1625. Alder-man 1665. Mayor 1693. A Burgess for Parliament. Died 1679. | Others. | Francis. = Ann. | Thomas. Died in 1632. |

| Thomas, Mayor of Reading, = Eleanor 1699-1712. Burgess for Parliament. | | | Francis, Mayor of Reading = Rebecca. 1689-90. Burgess for Parliament. Executor of his mother's Will. | | | Died 1699. | | Timothy of Reading. = Elizabeth. Born 1682. | Elizabeth. Born 1684. |

Cordelia. = John Hollidge, of Bristol. Born 1709. Died 1733.

Timothy. Born 1693. =

PEDIGREE OF SOME TYRRELLS OR TERRELLS, AND COLLATERALS, OF AMERICA

CHART 4.

Robert Tyrrell, of Reading (Chart 2) = Jane, dau. of Robert Baldwin, 29 June, 1617, at St. Giles' Chettb, Reading

| John. Born 26 June, 1618. Will 1661 | Robert. Born 14 November, 1619. Will 1677 | Richmond, of New Kent, Va., in 1668 = | William, of New Kent, Va., in 1687. = Martha | Mary = John Mewe. Born at Reading 2 April, 1620 |

William, of Hanover Co., Va. Died 1727. = Susannah Watson.   Mary, baptised at St. Mary's, Reading, 24 January, 1654

| William. | Timothy, of New Kent, Va. Born 1688 | = Elizabeth, dau. of John Foster | Joel. = | John. | David. = Agatha Chiles. (See page 28.) Died 1757 | Henry. = Ann Chiles. Died 1760. = 2 Sarah Woodson. (See page 28.) | James. Died 1734. | Ann = David Lewis in 1717 (See page 29.) |

| Robert, of Orange Co., Va., 1697-1786 = 1 Mary Foster. = 2 Judith Towles. | Joseph | Joel. | Elizabeth | Mary | Ann. | Henry (Major) = Ann Dabney | Joel = Anne Lewis. | Dau. = Stephen Willis. |

| Robert. | William = | John = Elizabeth Towles. | Edmond, of Orange Co., Va., 1743-84 = Margaret, dau. of Col. Henry Willis and his wife Mildred Washington, aunt of the President | Ann. Sarah Mary. | Richmond, = Cecilia Derracott, of Virginia | Joel = Martha Williams. | William-Gaylord. William-Lewis. Peter-Higgins. Mary. Anna. Susanna. Frances. Jane. | Stephen = Martha Willis (Colonel) 1780-1850. Wharey. |

| Son. = | Oliver, = Susan, dau. of Uriel Mallory | John, of = Rebecca Howard Cornelius 1783-1830 | Others | Mary-Foster. = Reuben Rucker, Born 1760 of Virginia in 1785 | Nancy Rucker, = John Henderson, 1789-1883 of Boone Co., Ky | George B. Willis = Mary Gallaher. (Captain) 1801-84 |

| Dr. George. | Uriel, M.D. = Jane, dau. of William Lovell | Margaret, = Allan Peage. 1790-1805 Died 1840 | James, = Susan 1797-1865 Mauning Cave. | Others. | William Henderson. = Martha Ann Paul, Born in Alabama. of Indiana. 1820-92 | Vespera Montalta = Alfred Freeman. Willis 1846-1902 1847-1920 |

| William Lovell, of Walnut Grove, Charlestown, Va. Died 1893. | Julia, dau. of John Tyler | Allan Peage. = Susan Catherine 1822-56 Frame 1854-1905 | Penelope = Joseph 1837-82 Picklin Cal. Born 1878 | Sarah Henderson. = Joseph Pyle Wiggins, Born 1847 of Indianapolis | Willie Freeman. Born 1884 |

| Julia Lovell Terrill, of Walnut Grove. | Rachael Margaret = Ancil Clayton Moody Peage. 1845-1907 Born 1845 | Walter H. Picklin, = Mabel Rowlett of University Park, Kenosair Cal. Born 1878 | Dudley Howard Wiggins. Born 1871 |

| Alta Moody. Born 1897 | = Charles D. Babb, of Homer, Ills. | Joseph Kenneair Picklin Born 1905. |

Margaret Elizabeth Babb. Born 1899

PEDIGREE OF SOME TYRRELLS OR TERRELLS, AND COLLATERALS, OF AMERICA

(CONTINUED)

CHART 4 (Continued)

William Terrell, of Hanover Co., Va — Susannah Waters
(See page 26)

David Terrell, — Agatha Chiles   Henry Terrell — 1 Ann Chiles   Others   Ann — David Lewis in 1717
Died 1729   Died 1760   — 2 Sarah Woodson   Died 1734   1682–1779

David — Sarah   Henry, of Caroline — Mary, dau. of Captain   Thomas — Rebecca   Ann — Charles   Betty. — Zachariah   Hannah Lewis. — James Hickman
1720–1808   Johnson   Co., Va   William Tyler   Born   1736–1804   Pentross   Lynch   Morman   1728–1853   1724–1816
    1735–1812   1749   1744

Edward — 1 Mary Johnson   John, of Spotsylvania — Abigail, dau. of   Joseph — Sallie Terrell   David Hickman — Clara McClanahan
Born 1753 | — 2 Jane Johnson   Co., Va 1772–1811 | Archibald Allen   Born 1777   Born 1749   Died 1820

John Harrison — Sally Moore.   Others   Williamson, of Clark Co., — Martha, dau. of Joseph   Joseph. — Mary Anderson   David McC. Hickman — Eliza Keller
1808–67   Ky. 1805–73   Terrell. Born 1808.   1813–73   (Captain) Born 1788   Johnston
                                                                                 1815

Lynch Moore — Mattie   William Henry — Eliza   Edwin Holland, of San Antonio, — 1 Mary Maverick.   Charles Milton — Salla   Others   Hon David H — Anna C. Bryan
(Colonel), of   Hammond   Harrison   Church   Texas 1843–1910   Sometimes — 2 Lala Lasater   (General)   Speake.   Hickman
Atlanta. Born   (General)   U S Minister to Belgium   1825–1904   1821–68
1864   1929–84   1861

French Jay, — Rosa   Lisa Severson   George Pisher —   Maverick   Lewis   Lasater   William   Henry   Mary D Hickman. — John Ewing
Born 1875 | Robinson   William Henry   1865–97   George   Martha.   Mary   Frederick   Cordelia   Born 1866   Price
             Harrison   Dorothy   1895
             Norwood Lynch   Hickman Price.   Andrew Price.
                                                                          Born 1886   Born 1890.

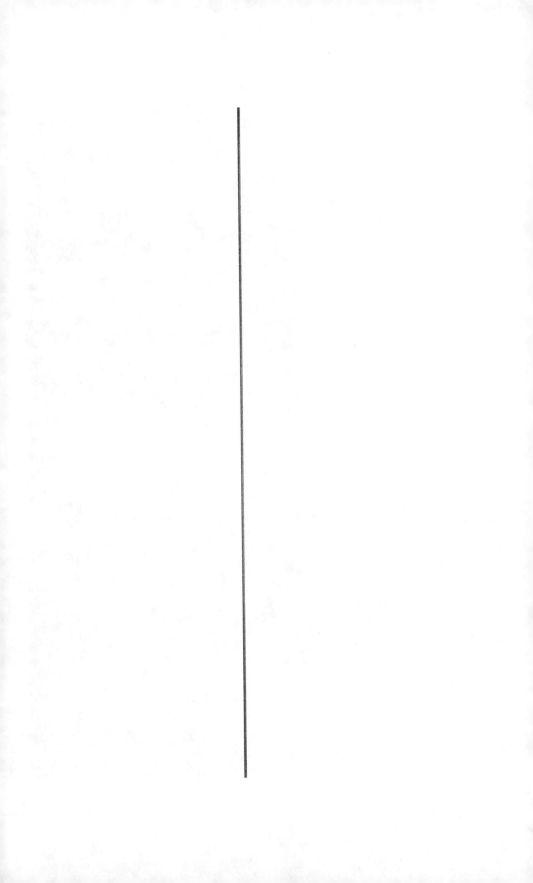

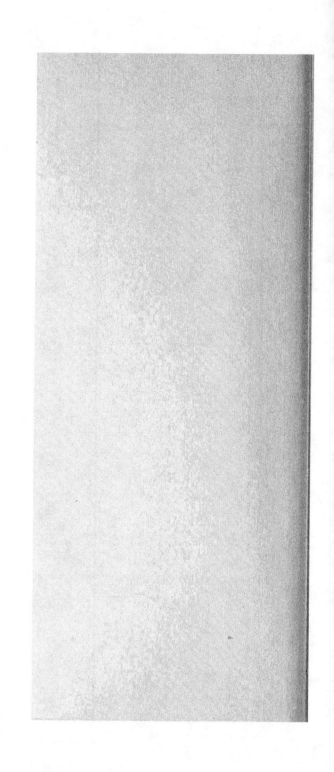

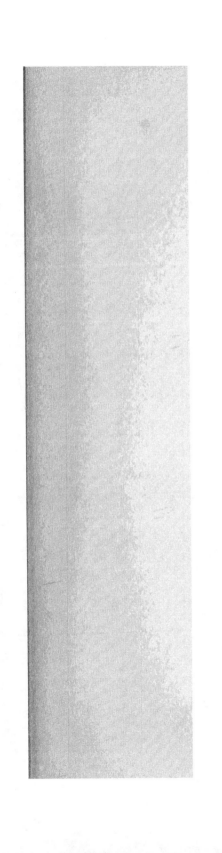

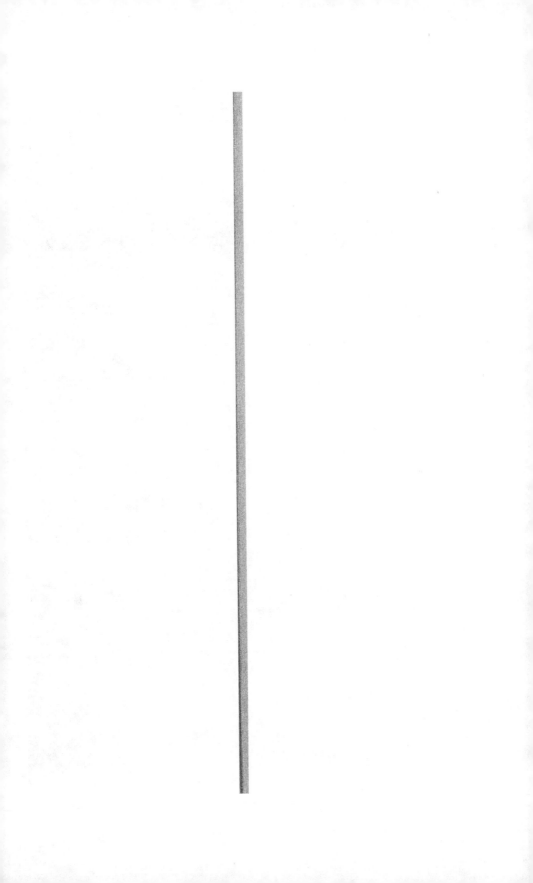

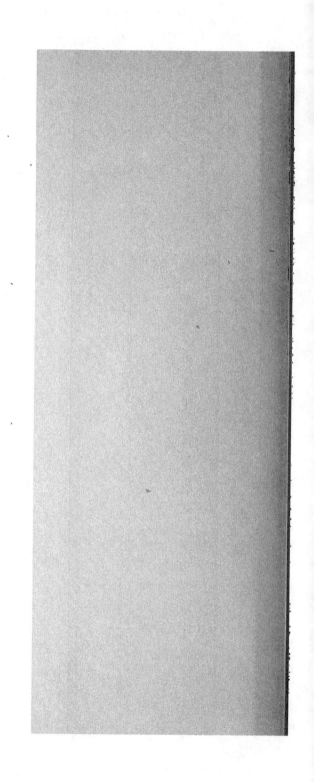

Printed in the USA
CPSIA information can be obtained
at www.ICGtesting.com
LVHW012205251123
764915LV00004B/83